Amalia Aventurin

Petroleum Geochemistry

GRIN Verlag

Bibliografische Information der Deutschen Nationalbibliothek:

Die Deutsche Bibliothek verzeichnet diese Publikation in der Deutschen National-
bibliografie; detaillierte bibliografische Daten sind im Internet über http://dnb.d-
nb.de/ abrufbar.

Imprint:

Copyright © 2014 GRIN Verlag GmbH
Druck und Bindung: Books on Demand GmbH, Norderstedt Germany
ISBN: 978-3-656-72257-1

This book at GRIN:

http://www.grin.com/en/e-book/278498/petroleum-geochemistry

Practical Laboratory Class

Petroleum Geochemistry

Winter Term 2013/2014

Content:

Petroleum Geochemistry

I. Summary

Eight different samples of various locations have been analysed during the practical course. The samples consisted of two source rock samples (Kimmeridge Clay and Blue Lias), two oil sands (Wealden oil sand and Osmington Mills oil sand) and four oil samples including Libya oil, R828, Iraq oil (09_205) and Schwedeneck. The Kimmeridge Clay represents an immature source rock with very good potential. The Blue Lias source rock, kerogen type II, shows the same conditions as the Kimmeridge Clay. The other Kimmeridge Clay sample kimsrf 1 indicates kerogen type I. The Wealden oil sand shows intense biodegradation and plots in the area of a type II kerogen. The Osmington Mills oil sand plots in type I kerogen area, and also shows intense biodegradation. The Libya oil is comprised of a type II/III kerogen, similar to the R828 oil sample which additionally shows oxidizing marine conditions. Iraq oil plots in the area of a type II kerogen and was deposited under marine hypersaline conditions. The last sample, the Schwedeneck oil, was deposited under rather oxidizing conditions and sterane analysis suggests an early oil window stage.

II. Introduction

The aim of the laboratory course „Petroleum Geochemistry" in the winter term 2013/2014 was to assess the kerogen type, depositional environment, maturity and petroleum potential of source rock and oil sand samples, using standard geochemical screening and characterization methods. The experimental procedures were based on the Norwegian Industry Guide to Organic Geochemical Analyses (NIGOGA). They comprise screening methods on bulk source rocks such as TOC-measurements and Rock-Eval Pyrolysis, as well as semi-quantitative analysis of bitumen composition by means of chromatographic separation. The fractions gained by liquid column chromatography (saturated, aromatic, NSO) were further characterized by GC and GC-MS measurements, allowing the recognition and quantification of biomarker compounds. The schematic work flow according to the NIGOGA guide is depicted in figure 1 (Appendix).

The set of samples (see Table 1) comprises two Source Rocks and two Oil Sands, all collected during the field trip "Reservoir Petrology" in South England in September 2013. The exact locations where the samples were found are shown in figure 2 (Appendix). For the GC and GC-MS analysis some of the samples were substituted by those from the year before because in those the biomarkers were better preserved. In addition, three oils from different locations were examined. Again, for the GC-MS analysis an additional Oil from Schwedeneck was analyzed because it was suitable for exhaustive biomarker analysis.

No.	Type	Formation	Location/Outcrop
1	Source Rock	Blue Lias	Lyme Regis
2	Source Rock	Kimmeridge Clay	Kimmeridge Bay
3	Oil sand	Wealden	Lulworth Cove
4	Oil sand	Corallian	Osmington Mills
5	Oil		Lybia
6	Oil		R828
7	Oil		Iraq (09_205)
8	Oil		Schwwedeneck

Tab.1: Analyzed sample set, comprising two source rocks, two oil sands and four oils.

III. Results and Discussion

i. TOC

The TOC results of the samples vary between L.R. (Lyme Regis/Blue Lias), KC (Kimmeridge Clay) and the oil sand (O.S.) samples. L.R. has around 10 % TOC, which is remarkably high and has been measured twice, to clarify this value. The KC source rock shows also a high TOC result, but has a lower value than the L.R. source rock. Finally, the O.M.OS (Osmington Mills/Corallian Oil Sand) and WOS (Wealden Oil Sand) samples show lower TOC results. Their comparable origin from oil sands builds a separated category, at least for TOC.

Sample	Weight in [mg]	TOC in [%]	TIC in [%]	TC in [%]
O.M.OS	100.4	1.839	0.1111	1.95
L.R.	100.1	9.55	2.677	13.23
L.R. II	100.5	9.535	2.756	13.29
KC	100.1	6.868	0.3817	7.25
WOS	100.3	2.044	0.0844	2.13

Tab.2: Results of the TOC, TIC and TC measurements of source rocks and oil sands.

(Note: L.R. & L.R. II are two measurements of the same sample, to clarify the high TOC/TIC content)

ii. Rock Eval

For the Kimmeridge Clay (KC) source rock, the Wealden oil sand (WOS), the Corallian oil sand (OM) and the Blue Lias (BL) source rock sample a *Rock-Eval Pyrolysis* was performed. Measurements include temperatures over time, S1 and S2 in mV (with FID) and the CO_2 content in mV (with IR). Through a slightly differing Rock-Eval assembly and a temperature measurement in the stemp, 43 °C have to be subtracted in order to get correct T_{max} temperature values. The calculations were performed with two standards (STD).

Sample	Weight	TOC	S1	S2	S3	Tmax
#	[mg]	%	[mg/g]	[mg/g]	[mg/g]	[°C]
Blue Lias	50,6	9,55	0,41	62,47	3,49	411
OM 1	100	1,839	7,73	14,14	1,19	832
OM 2	100,8	1,839	7,13	14,14	1,24	832
Standard 1	100	7,65	2,55	50,75	1,91	430
WOS	98,9	2,044	7,86	13,54	1,43	832
KC 1	100,5	6,868	1,28	44,88	1,00	417
KC 2	101,5	6,868	0,02	0,08	1,00	832
Standard 2	99,8	7,65	2,49	51,48	1,51	429

Sample	PI	HI	OI	PY	Kerogen Type	R.C.I
Blue Lias	0,007	654,136	36,545	62,88	17,900	65,843
OM 1	0,353	768,896	64,709	21,87	11,882	118,923
OM 2	0,335	768,896	67,428	21,27	11,403	115,661
Standard	0,048	663,399	24,967	53,300	26,571	69,673
WOS	0,367	662,586	70,044	21,408	9,460	104,736
KC 1	0,028	653,465	14,560	46,160	44,880	67,210
KC 2	0,200	1,165	14,560	0,100	0,080	0,146
Standard	0,046	672,900	19,714	53,968	34,132	70,547

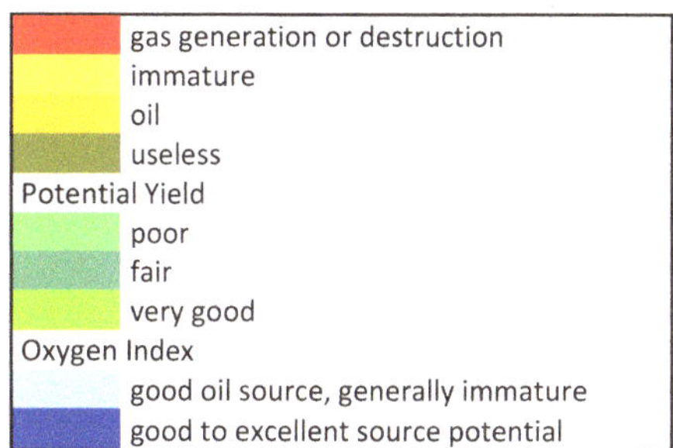

Fig. 1: Results of Rock Eval and calculation of necessary parameters showing their classification with color codes.

The Rock Eval analysis for the sample KC2, WOS, OM1 and OM2 did not conduct sufficient results due to T_{max} values of >800°C

Following equations were used for calculations:

$$\text{Potential Yield (PY)} = S_1 + S_2$$

$$\text{Production Index (PI)} = \frac{S_1}{(S_1 + S_2)}$$

$$\text{Hydrogen Index (HI)} = \frac{100 * S_2}{\text{TOC \%}}$$

$$\text{Oxygen Index (OI)} = \frac{100 * S_3}{\text{TOC \%}}$$

$$\text{Kerogen Type} = \frac{S_2}{S_3}$$

$$\text{Reactive Carbon Index (R. C. I.)} = \frac{(S_1 + S_2) * 10}{\text{TOC \%}}$$

Wealden (oil sand)

For the WOS sample the T_{max} value of 832°C is too high and therefore not important for further calculations. The results of the Production Index and Generation Index depict that the sample can be characterized as immature. Furthermore, the results of the Kerogen Type and Hydrogen Index indicate oil. This is supported by the Oxygen Index. The Potential Yield is with a value of 21 ranked as very good and the sample has a Reactive Carbon Index (R.C.I) of 104.73. If T_{max} is greater than 475°C or the Generation Index is greater than 0.6, the Reactive Carbon Index becomes less useful. Figure K shows that the S1 peak is greater than S2, indicating the oil window. The S3 peak is significantly smaller than the S1 and S2 peak, so the amount of CO_2 formed by thermal breakdown of kerogen is very low. The S2 value >5 mg/g is an indicator for an excellent source potential or high molecular weight soluble organic matter (biodegraded oil) or coal. The first peak is > 1 and therefore an indicator for large amounts of kerogen-derived bitumen or the presence of migrated hydrocarbons.

Kimmeridge Clay (source rock)

The sample of Kimmerdige Clay was analysed two times to have a duplicate. Similar to the Wealden oil sand, the KC1 sample is characterized as an immature oil, but it has a good to excellent source potential. Moreover it has a very good Potential Yield with a value of 45 and a R.C.I value of 66. Here the S2 is greater than the S1 peak and S3 is very small (Figure G and H). The S2 value is > 5 mg/g. S1 is >1.
The duplicate KC2 shows not useful values due to a T_{max} of 832 °C (Figure I).

Blue Lias (source rock)

The Blue Lias sample shows a maximum temperature of 411 °C and is therefore immature. This is supported by the Production Index and the Generation Index. The Hydrogen Index and the Kerogen Type indicate oil. The values of the Oxygen Index are high so they indicate a good to excellent source and the Potential Yield is classified as very good. The Reactive Carbon Index shows a value of 65.84. Figure E shows that the S1 and S2 peaks are identically that indicates a homogenous sample and the S3 peak is very small (Figure F). The value of S3 is 3.49 mg/g which indicates relative large amounts of terrigenous matter. The S2 peak has a value > 5 mg/g.

Corallian Osmington Mills (oil sand)

The sample of Osmington Mills was analysed two times to have a duplicate. Both depict a maximum temperature of above 800 °C, which is less useful for interpretation. The Production Index varies (0.353 and 0.335) but both values suggest oil, the same applies to the Kerogen Index. The Hydrogen Index indicates oil as well, but the results of the Generation Index show gas or destruction. The Reactive Carbon Index cannot be analysed because it is useless when the T_{max} is above 475 °C. The values of the Oxygen Index are high so they indicate a good to excellent source potential and the Potential Yield is classified as very good. Figures A and C show that the S2 peak is higher than the S1 peak, which is an indicator for an immature sample. The S3 peak is as well as in the Blue Lias sample very low (Figure B and D). The S2 peak has a value above 5 mg/g. The first peak is above 1.

Conducted charts of the Rock-Eval Pyrolysis are attached in the appendix. The data was plotted into a Pseudo-van-Krevelen diagram (Figure N) and a crossplot of S2 versus TOC (Figure M), which shows that nearly all samples plot between kerogen type I and II. WOS is near type II and Om is near to type I. According to Van-Krevelen the sample KC 1 is plotted on type I kerogen, but after Langford and Blanc-Valleron it is in the range of type II. The Blue Lias sample is defined as type II according to Langford and Blanc-Valleron but shows a type I-II after Van-Krevelen.

iii.   Bitumen Extraction

The fraction of the sample, which still stayed in solution with DCM after one day, is defined as the possible bitumen. After separating it from the residual by filtrating and evaporating the DCM was in a rotary evaporator, the yield of extraction could be determined by weighing the bitumen and relating it to the bulk sample mass. Unfortunately the Bitumen extraction could not be completed in the laboratory for the Kimmeridge Clay and the Wealden Oil sand due to experimental mishaps. The results for the Blue Lias Source Rock and the Corralian Oil Sand are listed in table 3.

Sample	Bitumen mass [g]	Bulk sample mass [g]	Fraction of bitumen [%]
Blue Lias Source Rock		5	
Corallian Oil Sand		5	

Tab.3: Yield from the bitumen extraction on the Blue Lias Source Rock and the Corallian Oil Sand.

iv.   Asphaltene Precipitation

Two samples (R828 and 09_205) are analysed with respect to their amount of asphaltenes. The results are listed in the table 4 below. The oil R828 contains 5.286 % asphaltenes, whereas the oil 09_205 contains 32.88 % asphaltenes and the Libya oil has an amount of 11.92 % asphaltenes.

Process	R828	09_205 (Iraq)	Lybia oil
Oil weight [g]	0.2516	0.250	0.250
Flask weight [g]	24.2274	27.1971	24.6108
Filter dry [g]	0.9062	0.8867	0.9116
Flask + Remains of asphaltenes [g]	24.2407	27.2793	24.847
Remaining asphaltenes [g]	0.0133	0.0822	0.0298

Tab.4: Table is showing the results of the asphaltene precipitation.

v. IATROSCAN (TLC-FID analysis)

After the Iatroscan measurement, the peaks were categorized into Saturated HC, Aromatic HC, NSO components and Asphaltenes. The comparison of the calculated peak-areas gives the relative composition of the oils. Table 5 shows the results.

The LEK205 oil sample shows a normal composition of a normal crude oil, while the R828 Oil sample and the Libya Oil have a higher content of Aromatic HC. Due to the high aromatic content of the Libya oil kerogen type I can be excluded because no biodegradation is present in the GC. All oil samples show depletion of NSO compounds (resins).

Sample	SAT		ARO		ASPH & NSO	
	Area	Fraction	Area	Fraction	Area	Fraction
R828 - rod 3	30,05	41,72%	41,7	57,89%	0,29	0,40%
R828 - rod 5	35,31	44,81%	39,23	49,79%	4,26	5,41%
R828 mean	**32,68**	**43,26%**	**40,46**	**53,84%**	**2,27**	**2,90%**
LEK205 - rod 6	45,77	31,35%	77,11	52,81%	23,13	15,84%
LEK205 - rod 8	49,70	29,46%	91,66	54,32%	27,37	16,22%
LEK205 - rod 9	44,98	28,46%	76,77	48,58%	36,28	22,96%
LE205 mean	**46,82**	**29,71%**	**81,85**	**51,94%**	**28,93**	**18,36%**
Libyan - rod 7	54,01	70,04%	21,62	28,04%	1,48	1,92%
Libyan - rod 9	57,34	62,73%	28,37	31,03%	5,71	6,24%
Libyan - rod 10	52,94	67,54%	18,47	23,56%	6,98	8,90%
Libyan mean	**54,76**	**66,77%**	**22,82**	**27,54%**	**4,72**	**5,69%**

Tab.5: The IATROSCAN results for R828, LEK205 and Lybia Oil. All samples were measured multiple times. The mean values are based on the most reliable data. The column area shows the measured peak areas. The fraction is the calculated relative composition of the oils.

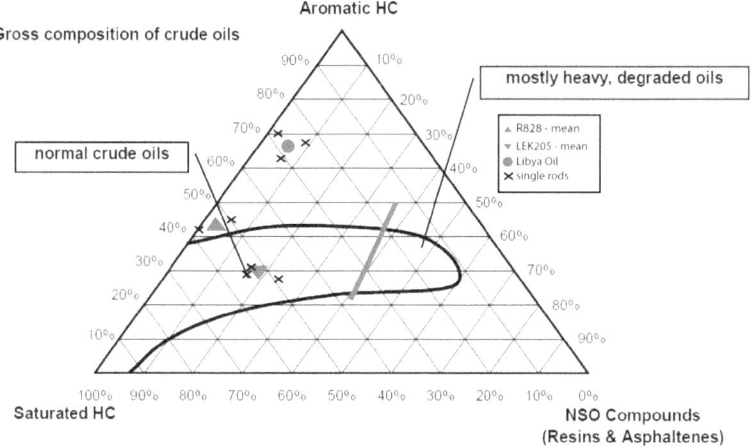

Fig.2: IATROSCAN plotted results. The triangles are indicating the mean values for LEK205, R828 and the circle for Lybia oil. The X's show single measurements.

vi.   GC (Gas Chromatography)

With the results of the gas chromatography and some suitable formulas, information can be given about depositional environment, maturity and possible biodegradation. All in all five different samples have been analysed.

The used formulas are as follows:

$$TAR = \frac{(nC27 + nC29 + nC31)}{(nC15 + nC17 + nC19)}$$

The Terrigeneous /Aquatic Ratio (TAR) gives information about the depositional environment, whether it is terrestrial or marine.

$$CPI = 2 * \left(\frac{(nC23 + nC25 + nC27 + nC29)}{nC22 + 2 * (nC24 + nC26 + nC28) + nC30}\right)$$

The Carbon Preference Index (CPI) gives information about the odd/even predominance and thus information about the level of maturity.

The pristane/phytane ratio shows, as well as the TAR, the depositional environment attached to the specific ranges:

>3 terrigeneous

<2 and >1 marine oxidizing conditions

<2 and <1 marine reducing conditions

<0.8 hypersaline conditions

The pristine/nC17 together with phytane/nC18 ratio gives information about depositional environment and level of maturity and biodegradation.

The calculated ratio derived from the results from the gas chromatography can be seen in Table 6. The interpretations which are derived from these ratios can be seen in Table 7.

Sample	TAR	CIP	Pr/Phy	Pr/nC17	Phy/nC18
Kimmeridge Clay	0.3544	2.4139	0.8513	1.6824	3.0341
Oil R828	0.0821	1.3006	1.5290	0.8459	0.6290
Lybia Oil	0.0460	1.0069	1.5545	0.4419	0.3833
Blue Lias	0.2970	2.3054	0.3589	1.5005	9.7926
Oil 09_205 (Iraq)	0.0415	0.9086	0.7484	0.1627	0.3192
Kimsrf1	0.3866	1.812	0.9394	0.9894	1.8333

Tab.6: The table shows the ratio calculated for the different parameters.

Sample	TAR	CPI	Pr/Phy	Pr/nC17 vs. Phy/nC18
Kimmeridge Clay	Marine	Odd predominance	Marine (reducing)	Type II kerogen, algal, marine strongly reducing
Oil R828	Marine	Odd predominance	Marine (oxidizing)	Type II-III kerogen mixtures, oxidizing
Libya oil	Marine	Odd-even equal	Marine (oxidizing)	Type II-III kerogen mixtures
Blue Lias	Marine	Odd predominance	Hypersaline	
Oil 09_205 (Iraq)	Marine	Even predominance	Hypersaline	Type II kerogen, algal, marine, strongly reducing (very mature)
Kimsrf1	Marine	Odd predominance	Marine (reducing)	Type II kerogen, algal, marine strongly reducing

Tab.7: Interpretation of the different samples based on the calculated values in Tab.6 (above).

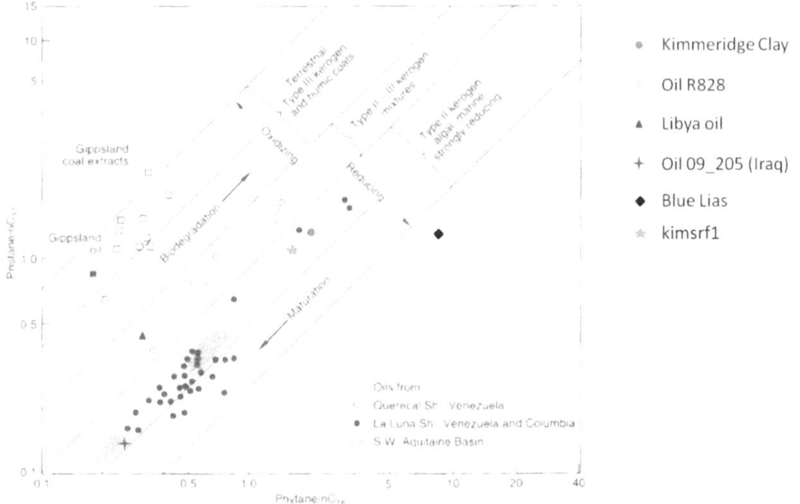

Fig.3: The figure shows where the different samples plot in the pristine/nC17 and phytane/nC18 diagram.

Kimmeridge Clay

TAR shows values for marine conditions as well as the pri/phy ratio. Additionally the pri/phy ratio indicates reducing conditions. The CPI has a high odd predominance with nearly 2.5 (Table 6). Figure 1 confirms these observations with Kimmeridge Clay plotting in the type 2 kerogen field with strongly reducing marine conditions. All these mentioned values do not speak for a high maturity.

Kimsrf1

The gas chromatography interpretation indicates an aquatic source (TAR: 0.387, nC27/nC17: 0.309), the CPI (1) ratio of 1.81 indicates a pre-, to early maturity. Reducing conditions associated to a type II or III kerogen and the early maturity are approved by the prystane-,phytane-, nC17- and nC18 values (pry/ph: 0.939, Plot of Phytane/nC18 vs. Prystane/nC17, fig. A).

Blue Lias

The TAR indicates a marine environment or more precisely as can be seen from the pristane/phytane ratio a hypersaline environment. The CPI shows high odd predominance, similar to the Kimmeridge Clay with nearly 2.5. High values for pristane/nC17 and phytane/nC18 makes a classification with respect to Figure 3 not possible, it plots outside of the classification. But the high phytane value indicates a low level of maturity (immaturity).

Oil R828

The TAR as well as the pristane/phytane ratio show marine oxidizing conditions. The CPI shows an odd predominance (1.3). In Figure 3 it plots type 2-3 kerogen mixtures area, with the lowest level of maturation is comparison to the other two oils.

Libya oil

This sample shows the same results for TAR and pristane/phytane like the Oil R828 (marine, oxidizing conditions). The CPI has a value of one, which shows that odd and even chain alkanes are in equilibrium. It plots in Figure 3 in the type 2-3 kerogen mixtures area but with a higher level of maturation then the oil R828.

Oil 09_205 (Iraq)

This sample indicates a marine (hypersaline) depositional environment. As the only sample the CPI shows even predominance. The pristane/nC17 to phytane/nC18 ratio (Figure 3) shows that it plots in the type 2 kerogen area with strongly reducing marine conditions. This sample is the most mature sample indicated by low pristane/nC17 and phytane/nC18 values.

vii.   GC-MS

Hopanes

In the Blue Lias sample the trisnorhopanes Ts and Tm, as well as hopane and C31-C34 homohopane were identified in the massfragment M/Z = 191. The calculated Ts/(Ts+Tm) ratio of 0.8 suggests a very high maturity. Furthermore the very high value suggests that a peak might have been contaminated with further chemical species. An evaluation of the homohopane index (HHI) was not possible, as C35 was not identified and the S and R hopanes were also not resolved.

For the Corallian oil sand sample the peaks for hopane, norhopane and trisnorhopane, as well as the C31 to C34 homohopanes were found. However, the trisnorhopanes could not be split into Ts and Tm. The C31 to C34 homohopanes could be split into S and R modifications. The calculated S/(S+R) ratios lie between 0.5 and 0.6 with a mean value of 0.5342, which suggests that the sample was close to reaching peak oil window conditions. However, the outcrop conditions at which the sample was taken suggests influence by water washing and biodegradation. Moreover the HHI could not be calculated due to the lack of C35.

For the Schwedeneck sample hopane, as well as the homohopanes C31-C35 were indentified. However, especially C35 must be treated carefully here because its value is close to the detection limit. Therefore the HHI was calculated at 0.05 suggesting a rather oxic depositional environment. Yet, this parameter is also influence heavily by maturation and possibly biodegradation. Ts/(Ts+Tm), as well as S/(S+R) homohopanes were not resolved and could therefore not be calculated.

Kimsrf1 maturity parameters 2S/22S+22R for C31 and C32 17 α hopanes (0.697); 17α hopane/ 17α Hopane+17β moretane C_{30} (0.85); 18 α neohopane/ 18α neohopane+17 α hopane C_{30} (0.692) were evaluated. Values correspond to a vitrinite reflectance range of 0.6Vr to 0.9 Vr (Table Y Fig. 14 and Fig.15 (Appendix)). The C35 Homohopane Index is associated with the redox potential in marine sediments. Higher values than 0.1 indicate anoxic conditions and are related to marine carbonates and evaporates. The heights of the homohopane peaks were evaluated manually, because of a better fitting of this scheme. The HHI is 0.083.

	Blue Lias (2013)	Corallian Oil Sand (2013)		Schwedeneck Oil	Kimsrf 1
	Source Rock	22S-config	22R-config		
C27 (Tm)	863889	1331537		-	235308
C27 (Ts)	3480225			-	367836
C29 (Norhopane)	-	2682024		-	28
C30 (Hopane)	2701272	3691916		3250146	101
C31	3769493	2030767	1390264	1881770	72
C32	3025466	1566720	1584311	499741	27
C33	1969018	1609228	1389451	230876	16
C34	776789	583313	561694	138309	11.5
C35	-	-	-	155142	10.5
HHI (C35/(C31-C35))	-	-		0.0534	0.083
Ts/(Ts+Tm)	0.8011	-		-	0.61
22S/(22S+22R)	-	0.5342		-	0.697

Tab.8: Peak areas of the most important hopane components and maturity parameters based on these biomarkers. At kimsrf 1 from C29 to C35 are the relative abundances given, but not the areas.

Steranes

The data gained from mass spectrometric analysis was processed with the help of the software *Excalibur* in order to identify and quantify the important biomarkers. Specific m/z ratios were extracted out of the mass spectrum and were plotted against retention time in mass chromatograms. This was done for the m/z ratios 372, 386 and 400, in order to identify cholestane, ergostane and stigmastane, respectively (See Fig. 20 to 22 (Appendix)). The integrated peak areas are listed in table 8. Where possible, different α/β and R/S configurations of the steranes were integrated separately. As maturity parameters serve the ββ-ratio and the 20S-ratio, which as well can be found in table 9, because these configurations have the highest stability. In addition, the relative proportions of C27, C28 and C29 can provide directions to the depositional environment.

	Blue Lias (2013)	Schwedeneck Oil	Kimsrf 1
C27 ααS	730301	1353557	1666852
C27 ββR	-	177859	2212689
C27 ββS	-	428406	2212689
C27 ααR	1664232	3872988	1041533
C28 ααS	940049	456133	1909349
C28 ββR	972967	740399	1555050
C28 ββS	1708276	377643	1555050
C28 ααR	724514	1965203	56151
C29 ααS	1620843	1086752	1343637
C29 ββR	-	203379	3225048
C29 ββS	1163440	501331	420988
C29 ααR	3687575	2947250	1340250
C29 [ββ/(ββ+αα)]	0.1798	0.1663	0.576
C28 [ββ/(ββ+αα)]	0.6170	0.3159	0.6128
C29 [20S/(20S+20R)]	0.4302	0.3351	0.59
C28 [20S/(20S+20R)]	0.6094	0.2356	0.496
C27 [%]	18	41	21
C28 [%]	33	25	29
C29 [%]	49	34	49

Tab.9: Peak areas of the most important sterane isomers and maturity parameters derived from them.

Unfortunately, steranes could not be identified in most of the samples. Except for the Schwedeneck oil they were either simply not present or have degraded already. In order to be able to analyze steranes on a source rock, a supplementary sample from the last year's practical course was provided (Blue Lias 2013). The [ββ/(ββ+αα)]-ratio of stigmastane is with 0.18 fairly low, indicating a low level of maturity. In contrast to this, the [20S/(20S+20R)]-ratio of C29, approximately 0.43, shows considerable degree of maturation. For comparison, these ratios were also calculated for ergostane yielding values >0.6 each, indicating that the sample Blue Lias (2013) is rather mature than immature. Despite the fact that the terrestrial C29 is the most abundant sterane, the Blue Lias sample can be classified as of marine origin according to the plot in figure 4. The Schwedeneck Oil undoubtedly shows low maturity indicated by the consistently small [ββ/(ββ+αα)]- and [20S/(20S+20R)]-ratios. Regarding the depositional environment no clear conclusion can be drawn from the relative abundances of steranes. The oil plots in the characteristic area for marine as well as lacustrine oils.

Steranes measured in the kimsrf1 sample are 20S/(20S+20R) C29 (0.49); and $14\,\beta\,17\,\beta\,/14\,\beta\,17\,\beta +14\,\alpha\,17\,\alpha$ C29 (0.576). The appropriate range of vitrinite reflectance of 0.55 Vr to 0.7 Vr is more narrow than the one of the hopanes. (Table Y Fig. 14 and Fig. 15 (Appendix))

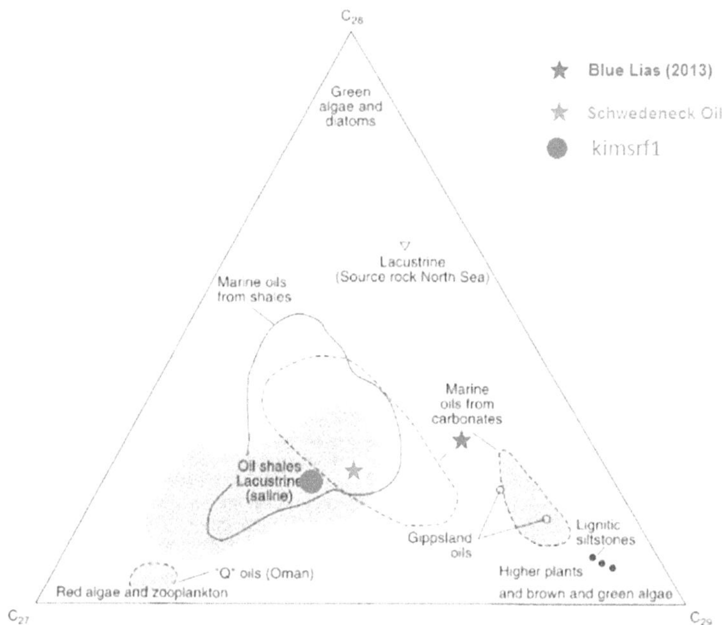

Fig.4: Depositional environment according to relative abundance of C27-C29 (after Peters et al, 2005).

IV. Conclusion

Blue Lias

The Blue Lias source rock with high TOC value of 9.55 % is a type II, marine kerogen. This is supported by the low TAR, as well as the Pr/Phy ratio displaying a hypersaline depositional environment. Moreover the sample is comparably immature, as suggested from T_{max} and Pr/nC17 and Phy/nC18. During GC-MS the steranes also suggest general immaturity which is only contradicted by the Ts/Ts+Tm ratio. The trisnorhopanes are the only parameter suggesting maturity and therefore it is assume that the sample is probably contaminated or influenced otherwise. Nevertheless the source rock shows a very good potential.

Kimmeridge Clay

The Kimmerdige Clay source rock shows a TOC value of 6.9 %, however kerogen type varies between type I and type II depending on the used plot. Due to the fact that only the van Krewelen plot suggests a type I kerogen, a type II marine kerogen as determined in the Langford & Blanc-Valleron, as well as GC measurements is applicable. Furthermore the reducing depositional environment was determined. Although the source rock is generally immature it shows a very good potential.

Kimsrf1

Summarized, the sample kimsrf1 is derived from an early mature kerogen with an appropriate vitrinite reflectance range of 0.5 Vr to 0.9 Vr and a mean of ca. 0.7 Vr. The sterane areas' derived parameters point to a more narrow region at 0.6 Vr to 0.7 Vr. These values are probably more reliable (Table Y, Fig.14 and Fig. 15 (Appendix)). The depositional environment was a marine or lacustrine and saline oil shale, either deposited while highly saline and reducing conditions predominated or with an input from higher plants (m/z 412, gammacerane, oleanane, Fig. 4 and Fig. 16 (only Fig. 16 in Appendix)). The homohopane index, slightly below 0.1 reasonably contrasts the hypothesis of a reducing environment: The depositional environment is not related to carbonates and evaporates, necessary for higher values (Peters et al. 2005). The gas chromatography interpretation confirms the aquatic source (TAR: 0.387, nC27/nC17: 0.309), the CPI (1) ratio of 1.81 confirms its pre-, to early maturity. Reducing conditions associated to a type II or III kerogen and the early maturity are approved by the prystane-,phytane-, nC17- and nC18 values (pry/ph: 0.939, Plot of Phytane/nC18 vs. Prystane/nC17, Fig. 3). Type III kerogens are excluded due to the confirmed aquatic regimen. A type II kerogen is possible, but a high input of higher plants in a marine environment is unlikely. The source of sample kimsrf1 is more probably a Type I kerogen accompanied by reducing, saline conditions

Wealden Oil Sand

The oil sand sample taken from a Wealden outcrop shows a TOC value of 2.04 % from a marine type II kerogen. Moreover, the only maturity information, which could be derived, is from the peak size in rock eval suggesting a rather mature sample. Biomarker determination was not possible, probably due to intense water washing or biodegradation at the outcrop.

Osmington Mills Oil Sand

The oil sand from Osmington Mills shows approximately 1.84 % TOC. Rock eval analysis determines a lacustrine type I kerogen, yet maturity information could only be derived from S/S+R steranes suggesting a high maturity, seeming logical as this is an oil sand. A value of approximately 0.55 for this was determined, thus, showing oil window maturity. GC analysis suggested intense biodegradation, as again probable due to exposed outcrop conditions.

Libya Oil

The Libyan oil sample is a crude oil, as determined in TLC-FID, comprised of a type II / III mixed kerogen. Moreover the GC analysis suggests an oxic depositional environment. Thus, biomarkers are maybe lost due to the depositional environment or maturity.

R828

The analysis done for the oil R828 included the TLC-FID measurement suggesting a crude oil. Furthermore the GC analysis showed a type II / III mixed kerogen under oxidizing conditions. Maturity information partially contradict each other, however the fact that this is an oil sample indicates a higher maturity. Biomarker analysis was not possible due to lack of resolution, which probably originates from the samples maturity.

Iraq Oil (LEK 205)

The oil sample from the Iraq showed a crude oil composition in the TLC-FID, which was deposited under marine hypersaline conditions. This type II kerogen is rather mature. Again biomarker analysis was not possible, probably due to maturity.

Schwedeneck

The Schwedeneck oil was only given for biomarker analysis. The HHI suggests a rather oxidizing depositional environment, however this must be treated with care seeing the C35 is close to detecting limitations. Furthermore the steranes suggest a rather early oil window stage.

However, the results from this practical course have to be treated with care due to possible failures during the experimental procedure.

V. References

The analysis and the report are based on 'The Norwegian Industry Guide to Organic Geochemical Analyses' (NIGOGA), Edition 4.0, 30th May 2000; published by Norsk Hydro, Statoil, Geolab Nor, SINTEF Petroleum Research and the Norwegian Petroleum Directorate.
Additionally the notes were used from the laboratory practical.
Peters, K. E., Walters, C. C., & Moldowan, J. M. (2005). The biomarker guide: Biomarkers and isotopes in the environment and human history (Vol. 1). Cambridge University Press.

VI. Experimental Procedures
i. TOC

During a TOC/TIC (total organic carbon/total inorganic carbon) measurement, the part of organic and inorganic carbon of a sample is measured in percentage. It is a standard measurement procedure in organic geochemistry. The measurement itself was done with solid state samples and used was a LiquiTOC II measurement instrument from "Elementare Analysengeräte GmbH". Oxygen is used as a carrier gas. Preparation of solid state samples is done with fine graded powder of a rock sample. For each measurement, 100 mg of a powdered sample is filled within a glass crucible and then placed in the instrument. The determination of the carbon peaks follows the procedure of a "three peak integration" method. Organic, as well as inorganic carbon is measured during the same measurement but in separated steps. The measurement is done with a "temperature ramp method". Here, the sample gets heated up to 550 °C and after that heated up to 850 °C. Each step takes around 7 minutes. During this time, organic carbon gets oxidized between 450 and 600 °C and inorganic carbon at around 800 °C. So both peaks have no intersection area and can be distinguished very clearly and precisely. One run takes about 28 min, including heating and cooling of the sample. Before a series of measurements is done, the instrument measures a standard with known TOC/TIC values. For our purposes we have used a 5.89 % TC (total carbon) standard. TC is the summation of TOC and TIC.

ii.   Rock Eval

The rock eval method is a bulk pyrolysis which is performed with a *rock eval 6* device. This method provides information about thermal maturity, kerogen type, thermal stability of kerogen, source rock, type of organic matter and migration. Different parameters are calculated, including production index [PI = S1/(S1+S2)], hydrogen index [HI = S2*100/(% TOC)], oxygen index [OI = S3*100/(%TOC)], maximum temperature [T_{max}], potential yield [PY = S1+S2], reactive carbon index [R.C.I. = (S1+S2)*10/(%TOC)], generation index [GI = S1/(S1+S2)] and kerogen type [KT = S2/S3]. The kerogen classification can be done with plots of HI versus OI and S2 versus TOC.

For the pyrolysis a small quantity of crushed sediment (<0.1g) is necessary which will be heated under a controlled temperature programming. The sample runs through a distinct time temperature program. At first the sample is heated to 300°C instantaneously for 3 minutes to detect the free hydrocarbons (bitumen) in mg HC / g rock, which is shown as the S1 peak. Subsequently the sample is heated at 25°C/min until 600°C, where hydrocarbon products are cracked from the kerogen (shown as the S2 peak) in mg HC / g rock. The last peak S3 shows the yield in CO_2 in mg CO_2 / g rock. The highest point of S2 furthermore shows the T_{max} value indicating the temperature of maximum hydrocarbon release. This value also increases with maturity.

iii.   Bitumen Extraction

In order to extract the bitumen from the two bulk rock samples, the origin rocks (K.C. and WOS) have to be crushed. There are two different steps necessary to do so. 10 g are weighed for each sample and is filled into an Erlenmeyer flask together with Dichlormethane (DCM) (100 ml) and copper (to reduce the sulfur in the sample) as well as two magnetic stirrers to mix it all. The Erlenmeyer flaks are put into an ultrasonic bath, where the samples are mechanically stirred for 15 min. If this process is finished the flasks are put to a magnetic plate where the mixture is stirred for another 30 min. The flasks were left in the fridge overnight.

The last step is to filter the samples. The samples were then separated from their rock remains by filtrating through a funnel. After this process the remaining solvent was evaporated out of the samples by using a rotary evaporator. At a temperature of around 40°C and 300 mbar to get rid of the DCM, which has a boiling point of 41°C at 1 bar. The solution was heated until only a solid fraction remained. Furthermore this solid fraction was then liquefied with 1 ml of hexane. With this prepared phase a fractionation can be done.

iv.   Asphaltene Precipitation

For the asphaltene precipitation around 250 mg of the oil from Libya were weighed inside a flask. Additionally 750 µl of DMC (CH_2Cl_2) were added to liquefy the sample. Furthermore 30 ml of pentane (C_5H_{12}) were added to precipitate the asphaltene compounds (high molecular weight hydrocarbons, NSO compounds), leaving the maltene compounds in solution. This flask was put into a refrigerator overnight. This step is mandatory to minimize coagulation of resins. While in the refrigerator, the maltenes and asphaltenes separate. Due to their higher molecular weights the precipitated asphaltenes sink to the bottom of the flask. In order to separate them completely the solution is run through a filter. Subsequently the asphaltenes (solid phase) remain in the filter and the maltenes (liquid phase) run through. The dried filter and wet are weighed. The amount of asphaltenes is 0.0298 g and this is determined by weighing the filter. The maltene fraction was then evaporated using a rotary evaporator until approximately 5 ml remained. This was done by heating at around 40 °C and 300 mbar, to get rid of the pentane which has a boiling point of 36 °C at 1 bar. Subsequently the solution was filled into a 5 ml flask and then remaining material was solved with 1 ml of pentane and also given into the 5 ml flask. Afterwards this flask was then further evaporated until approximately 1 ml remained. This 1 ml was then used for the column chromatography.

v.   IATROSCAN

The TLC-FID IATROSCAN is an automatic thin layer chromatography device, where the sample (mixed with DCM) is spotted on a so called chromarod, a glasrod, which is covered with silica gel, acting as the stationary phase and is then burned and afterwards detected in the FID (flame ionization detector). In order to do this several preparation steps are necessary.
At first 4 – 5 mg of the oil sample (Libya oil) was mixed with 0.5 ml DCM.
Around 2 µl of this mixture is spotted on three different chromarods (number 8, 9 and 10) and then it gets dried. Subsequently these are given into a bath of three different solvents. This started with a non-polar (n-hexane) solvent for 35 min, followed by a semi-polar solvent (toluene) for 15 min and lastly a polar solvent composed of 1 part DCM and 7 parts methane for 3min. After each step the chromarods dry on the air for 2 minutes. Due to capillary forces the solvent rises and takes along the parts of the kerogen, which are soluble in the used solvent. This leads to a succession, from top to bottom, of aliphatics, aromatics, NSO compounds and lastly leaving the asphaltenes, depending on their polarity. Afterwards the chromarods are placed in a drying oven for 90 s at 60°C. The last step is the actual measurement with the IATROSCAN MK-5 device. Here the chromarods are placed in the device and are scanned from top (more volatile) to bottom (non-polar fractions) (aliphatics to asphaltenes) this takes 30 sec. for one rod, a FID using a hydrogen flame. This signal is digitalized leading to a single peak for each component group.

vi.   GC

Gas chromatography analyses approximately 10-20% of all compounds, which have an appropriate volatility and consists of three main components. In the injector the liquid or gaseous sample can be injected (1µl). Subsequently, the carrier gas, which needs to be a high-purity inert gas (here: H_2), transports the sample into a column with a flow rate of 3 ml/minute. This column was a ZB-1 column with a length of 30 m, an inner diameter of 0.25 mm and a film thickness of 0.25 µm. Afterwards an electronic signal is generated in the detector which is based on the interaction of the solute with the detector. After the signal is processed it is plotted vs. the elapsed time, which is then called a chromatogram (fig. 6 to 13).

The whole process comprises that the sample is heated for three minutes at 60 °C. Furthermore it is heated up to 300 °C with 10 °C per minute. The volatile sample solutes are vaporized at 270 °C detector temperature (constant).

vii.   GC-MS

The GC-MS analysis is used to determine biomarkers in various samples. Through the coupling with a mass spectrometer the device allows the user to use "mass spectrometers" consisting of a chromatogram of a single mass. Subsequently through comparison one can evaluate the areas of the biomarker species.

The device used for GC-MS was a sector field mass spectrometer, which was coupled with a Hewlett Packard Series II 5890 chromatograph. A splitless injector was used leading to a Zebron ZB-1 silica capillary column (30m in length, 0,25mm width, 0,25µm silica). Furthermore helium was used as a carrier gas. In the oven a temperature ran through a program from 80 to 310 °C at 5 °C / min. At maximum temperature the conditions remained for 3 minutes.

Practical Laboratory Class

Petroleum Geochemistry

APPENDIX

Winter Term 2013/2014

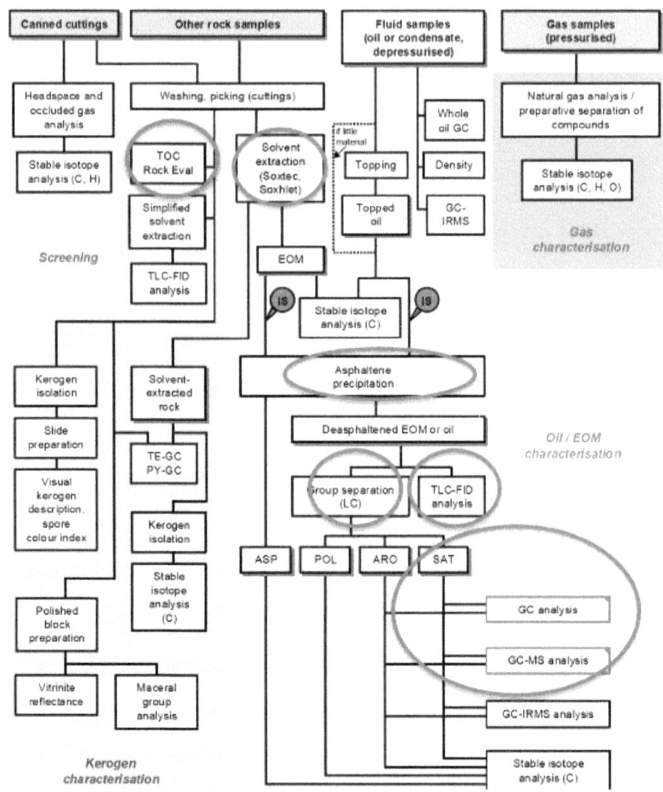

Fig.1: Oil and Source Rock characterization procedure according to the NIGOGA Guide.

Fig.2: Outcrop locations of the two source rocks and the two oil sands in South England

Fig.3: All fractionations and evaporated samples of Lybia oil.

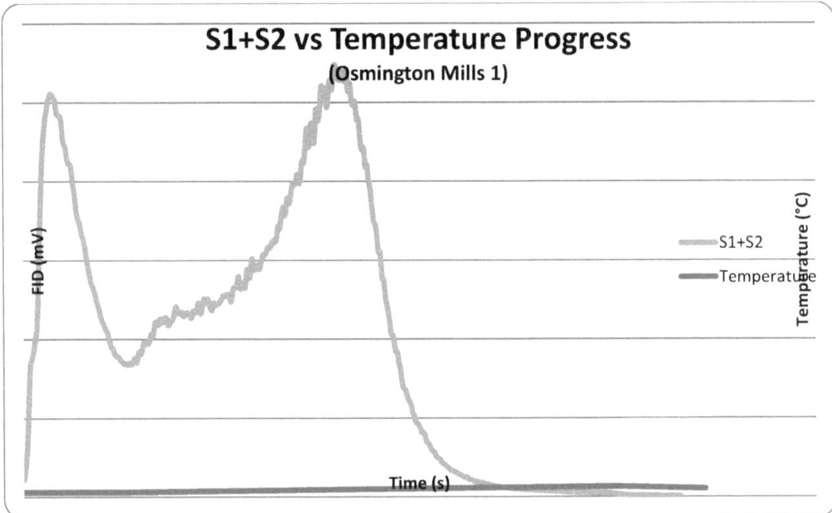

Fig.A: S1 &S2 peak vs temperature progress over time for Osmington Mills 1 (oil sand).

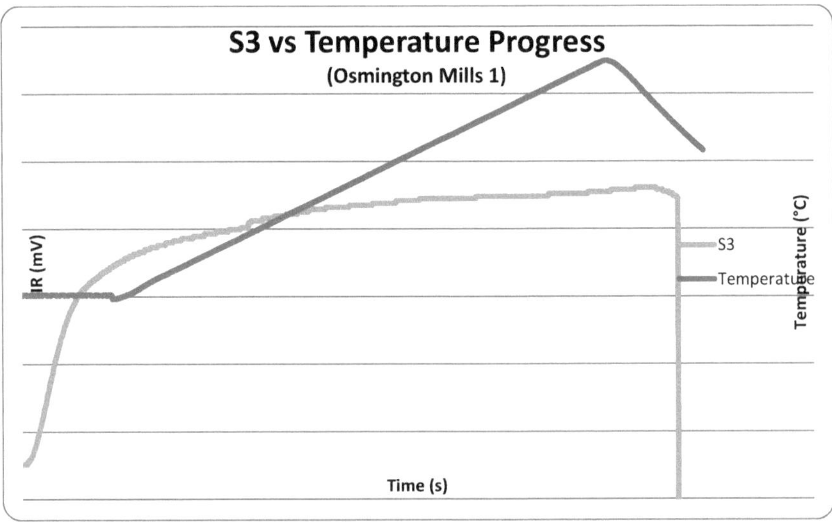

Fig.B: S3 peak vs temperature progress over time for Osmington Mills 1 (oil sand).

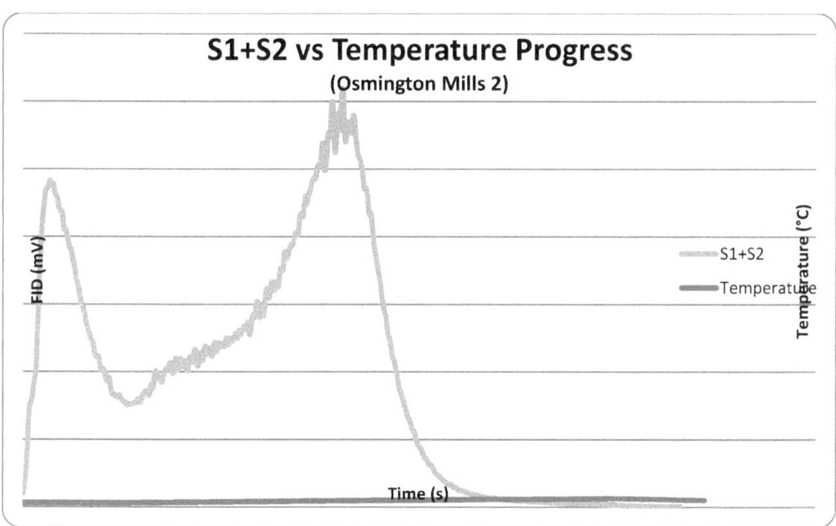

Fig.C: S1 & S2 peak vs temperature progress over time for Osmington Mills 2 (oil sand).

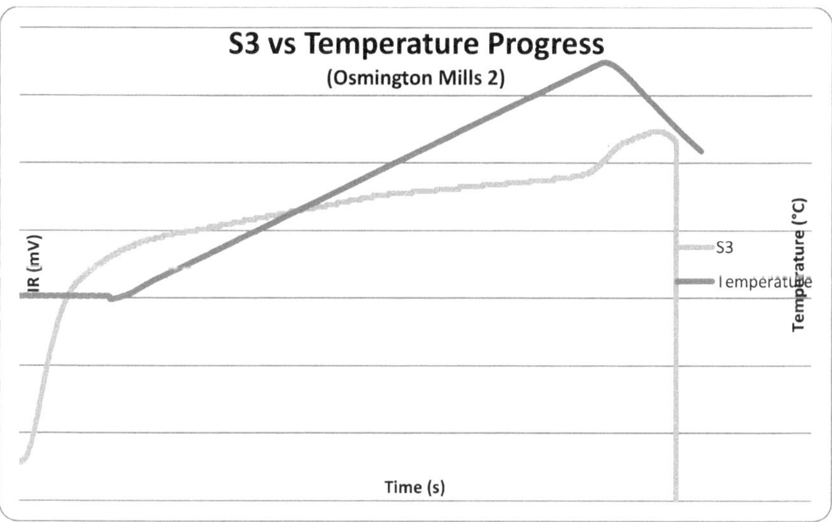

Fig.D: S3 peak vs temperature progress over time for Osmington Mills 2 (oil sand).

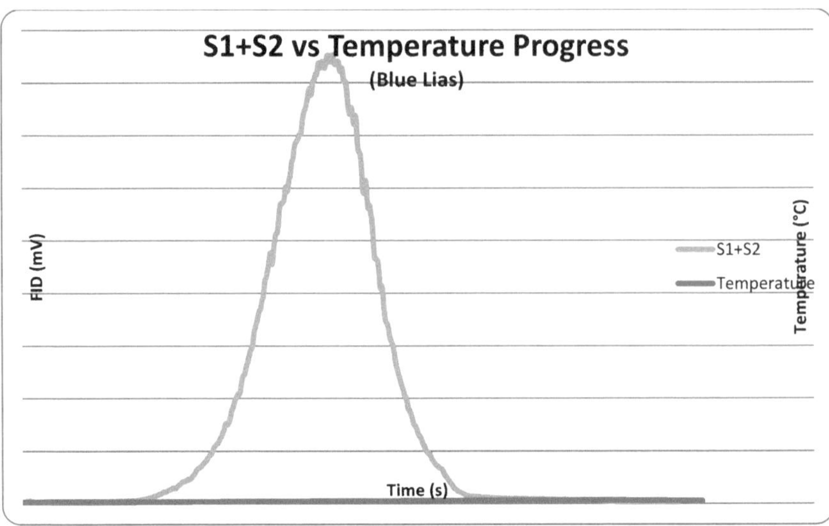

Fig.E: S1 & S2 peak vs temperature progress over time for Blue Lias (source rock).

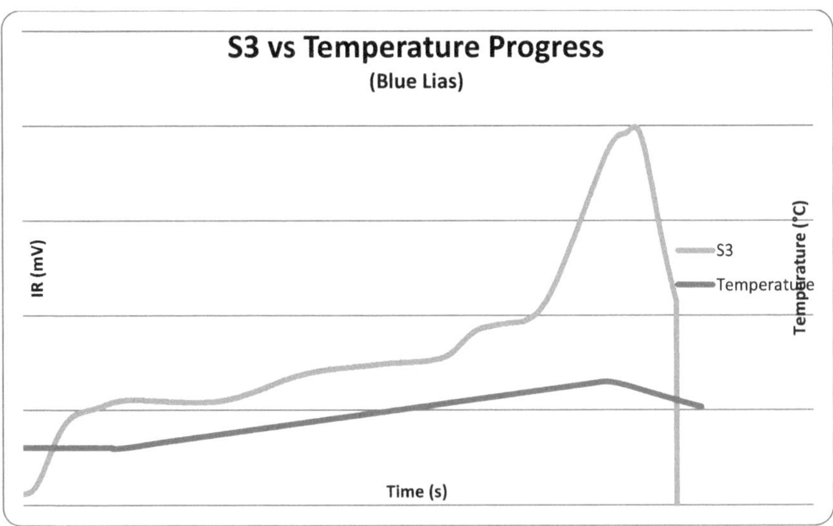

Fig.F: S3 peak vs temperature progress over time for Blue Lias (source rock).

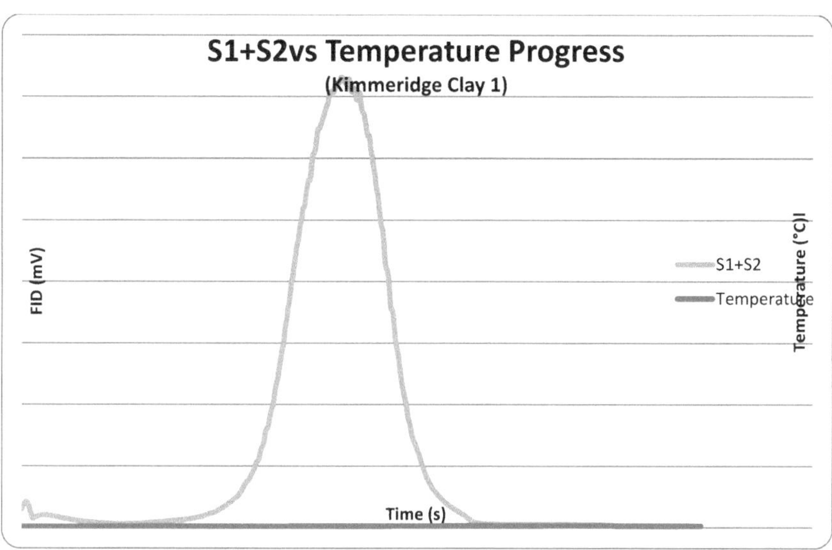

Fig.G: S1 & S2 peak vs temperature over time for Kimmeridge Clay 1 (source rock).

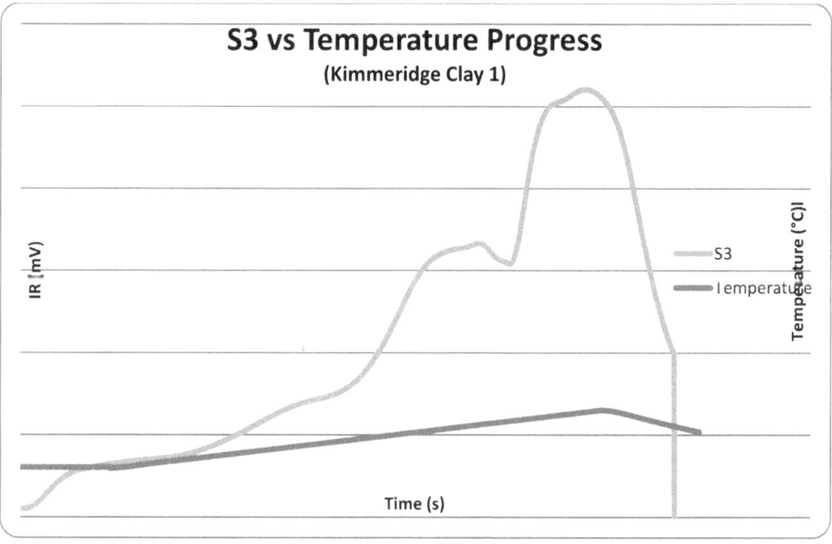

Fig.H: S3 peak vs temperature progress over time for Kimmeridge Clay 1 (source rock).

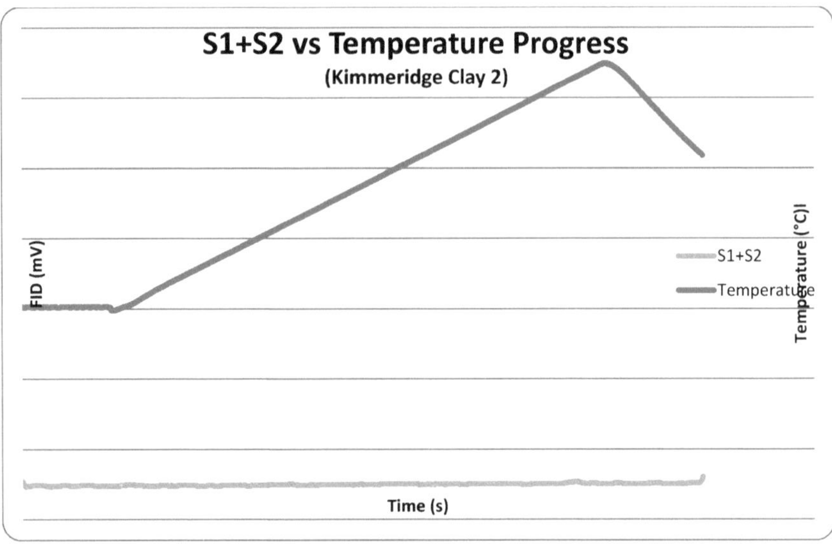

Fig.I: S3 peak vs temperature progress over time for Kimmeridge Clay 2 (source rock).

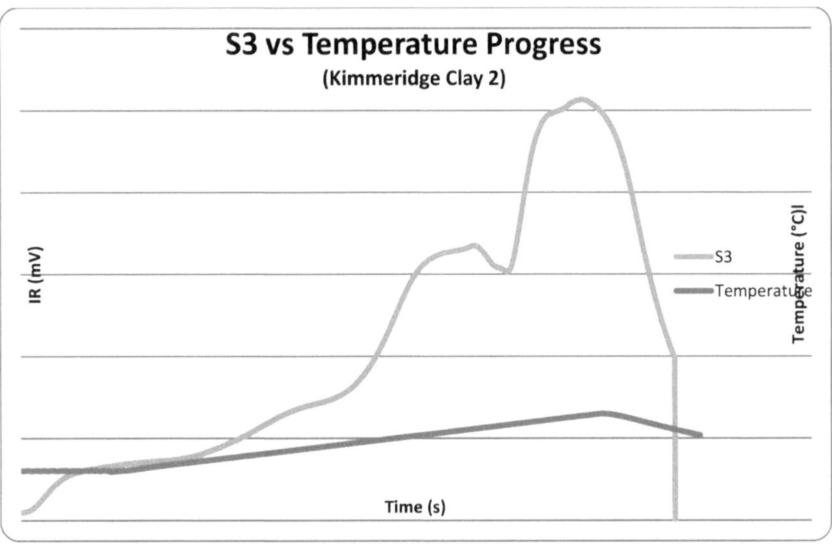

Fig.J: S3 peak vs temperature progress over time for Kimmeridge Clay 2 (source rock).

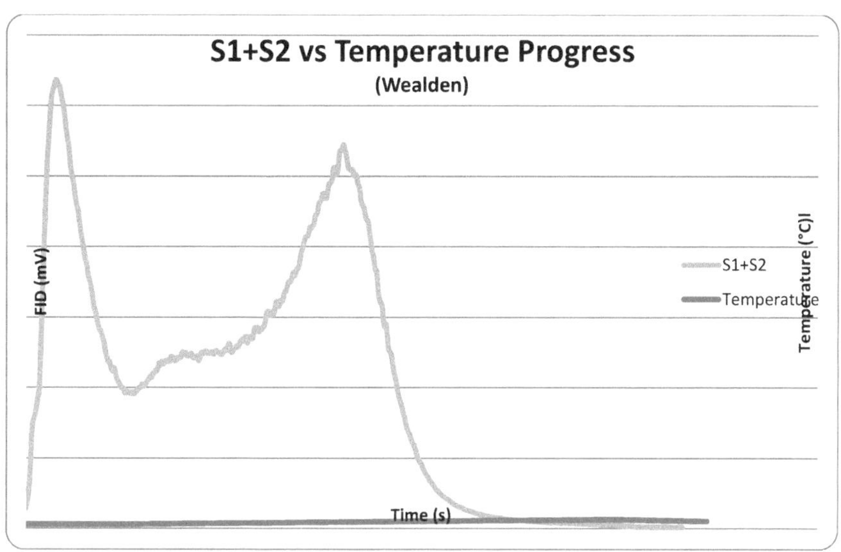

Fig.K: S1 & S2 peak vs temperature progress over time for Wealden (oil sand).

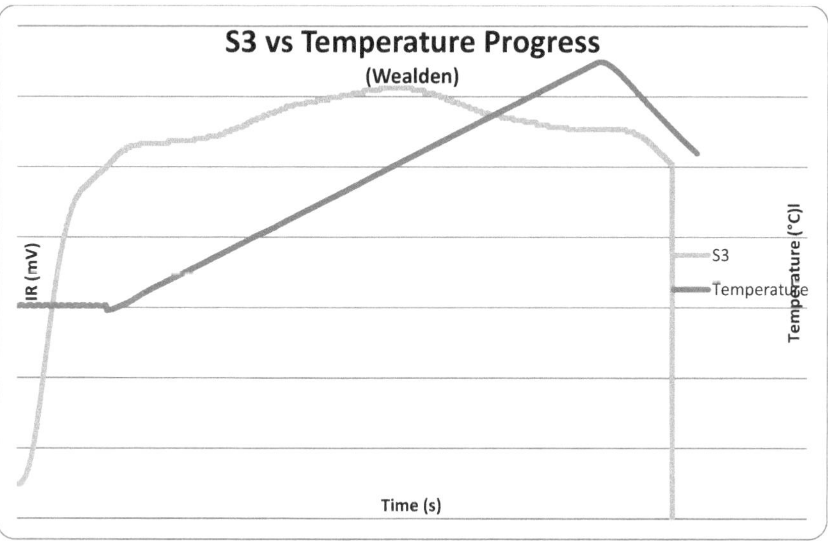

Fig.L: S3 peak vs temperature progress over time for Wealden (oil sand).

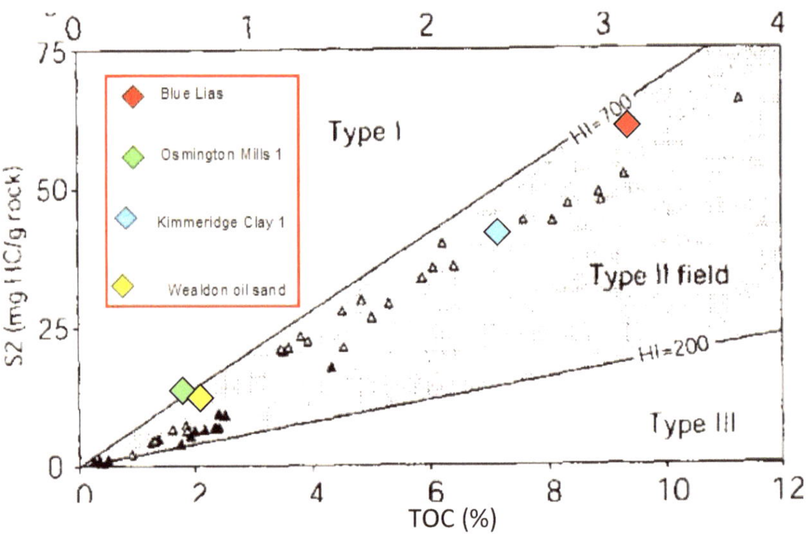

Fig.M: Crossplot of S2 (mg HC/ g rock) vs TOC (%wt) defines kerogen type of Kimmeridge Clay (after Langford and Blanc-Valleron,1990).

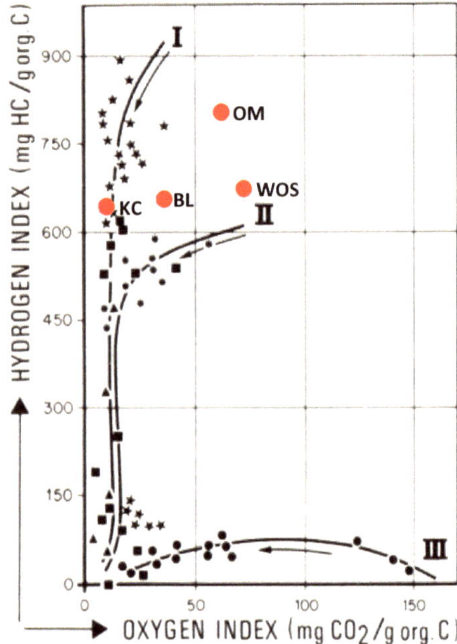

Fig.N: Crossplot of Hydrogen Index (HI) vs Oxygen Index (OI) defining kerogen type for Osmington Mills, Kimmeridge Clay, Blue Lias and Wealden (after van Krevelen).

IATROSCAN

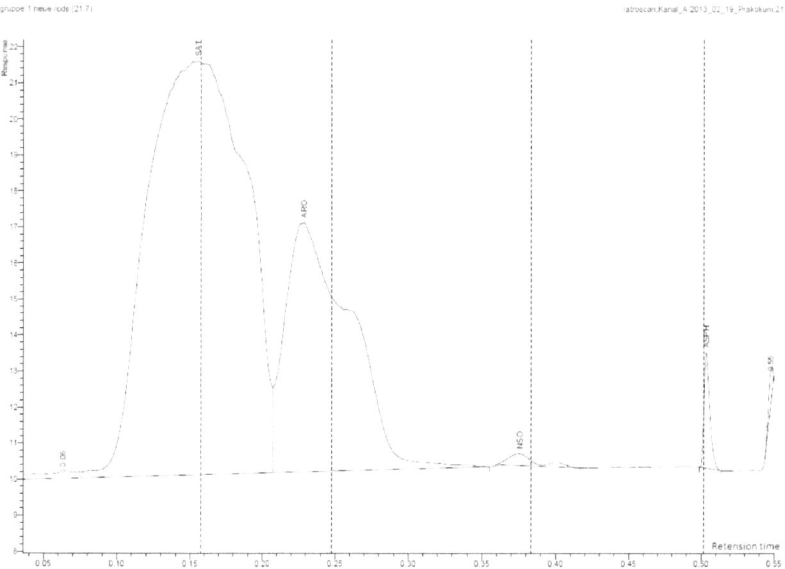

Fig.4: IATROSCAN diagram of the Lybia oil; rod no.7.

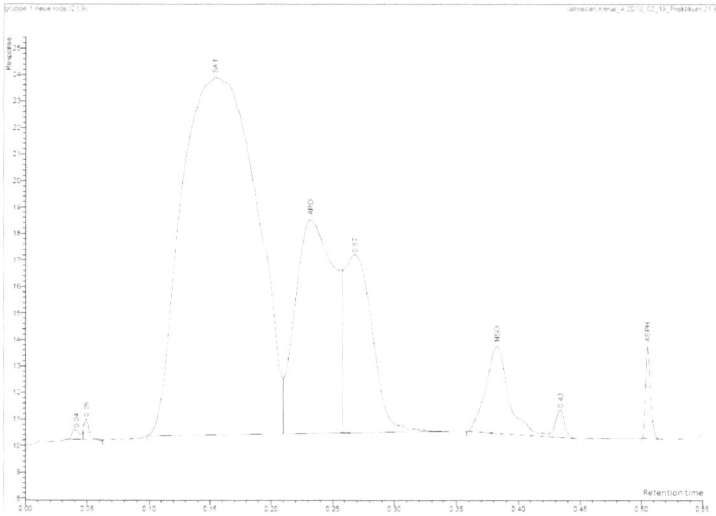

Fig.5: IATROSCAN diagram of the Lybia oil; rod no.9.

Gas Chromatogram (GC)

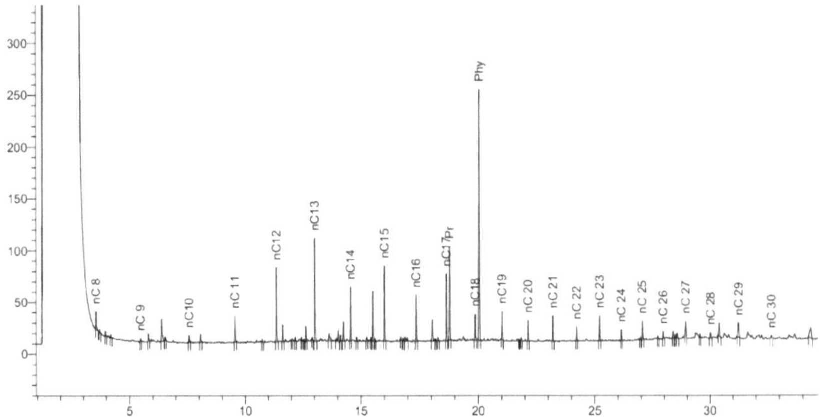

Fig.6: Gas Chromatogram (GC) of the Blue Lias Source Rock.

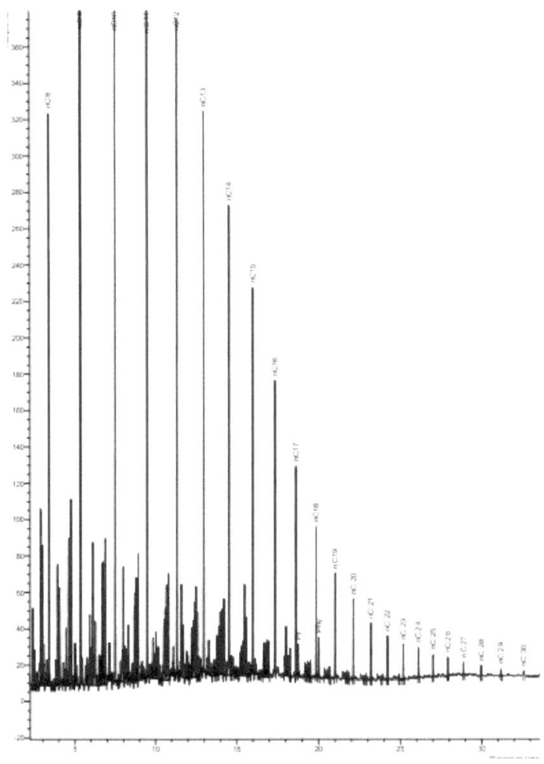

Fig.7: GC of the oil sample 09_205 (Iraq).

31

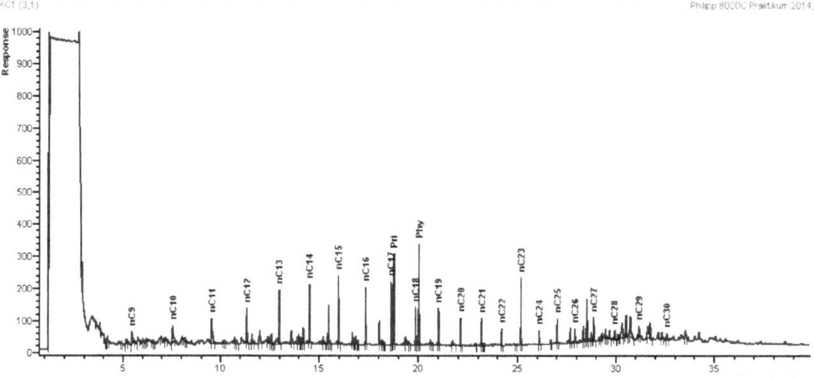

Fig.8: GC of the Kimmeridge Clay 1.

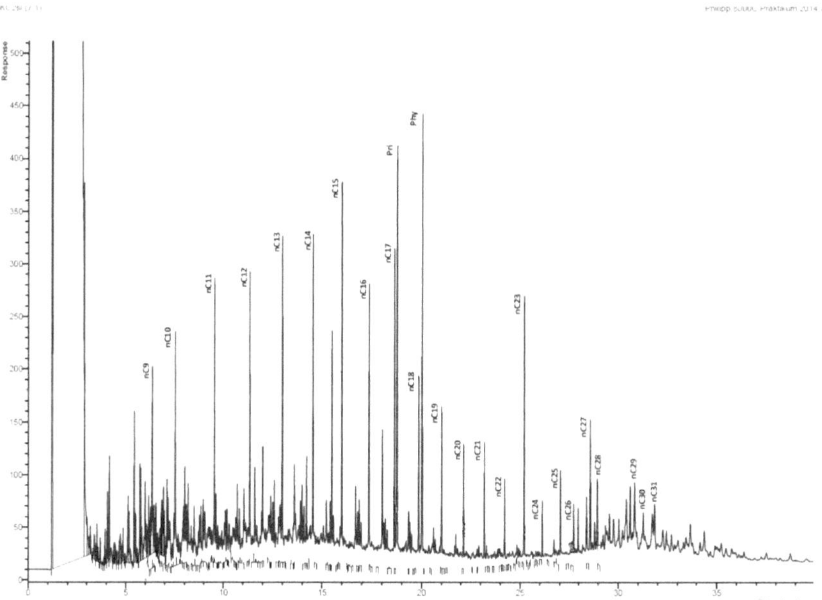

Fig.9: GC of the Kimmeridge Clay 2.

32

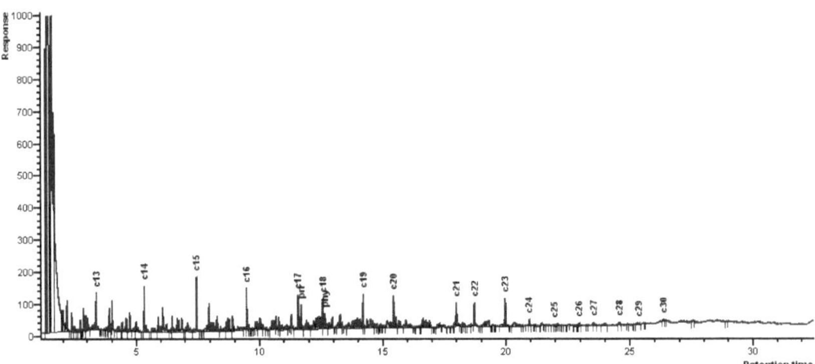

Fig.10: GC of the oil sample R828.

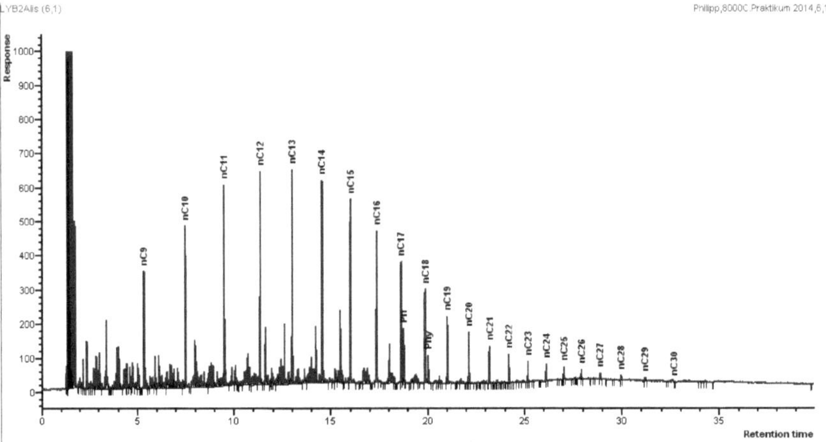

Fig.11: GC of the Lybia oil.

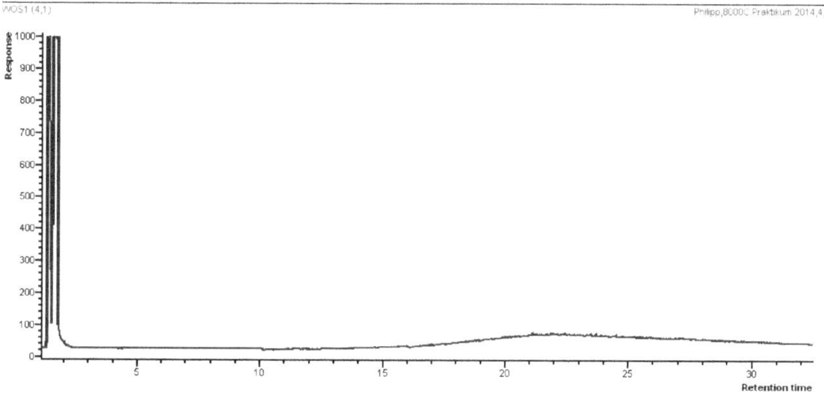

Fig.12: GC of the Wealden Oil Sand 1.

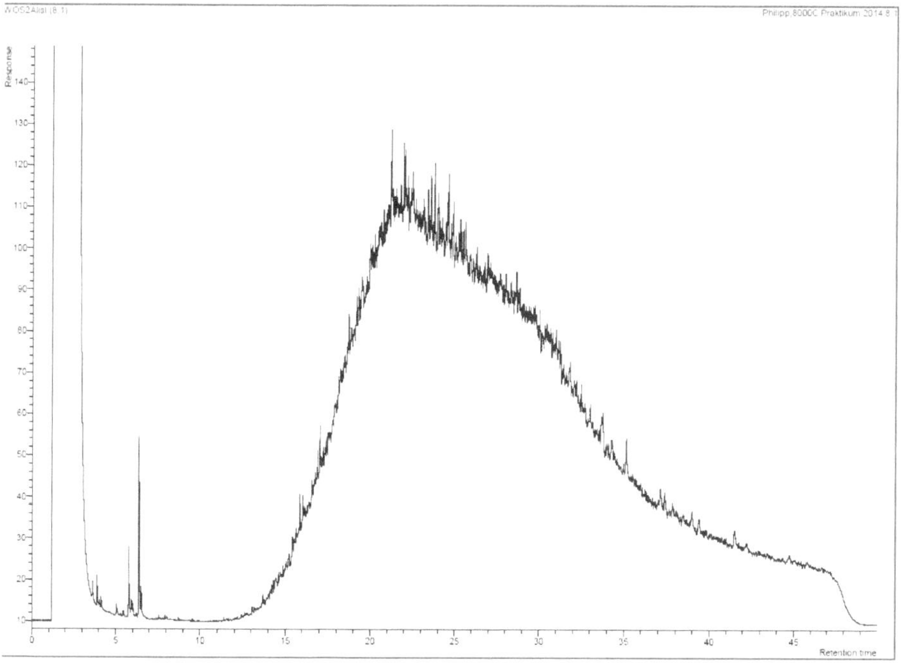

Fig.13: GC of the Wealden Oil Sand 2.

GC-MS

Kimsrf1

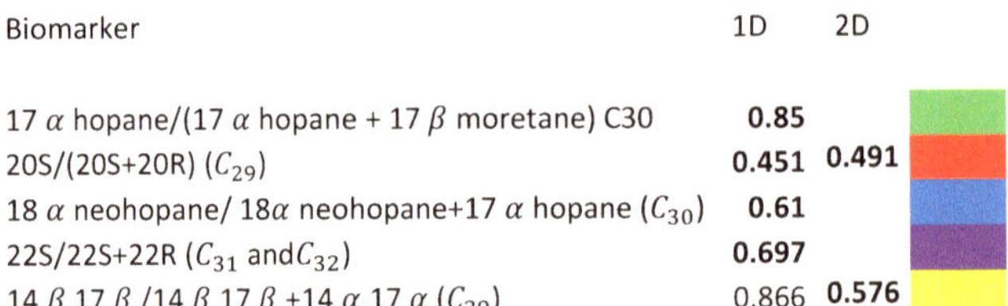

Biomarker	1D	2D
17 α hopane/(17 α hopane + 17 β moretane) C30	**0.85**	
20S/(20S+20R) (C_{29})	**0.451**	**0.491**
18 α neohopane/ 18α neohopane+17 α hopane (C_{30})	**0.61**	
22S/22S+22R (C_{31} and C_{32})	**0.697**	
14 β 17 β /14 β 17 β +14 α 17 α (C_{29})	0.866	**0.576**

Tab. Y: Sterane and Hopane maturity parameters. Colors indicate the placement of the bold values in figure 8 and 9.

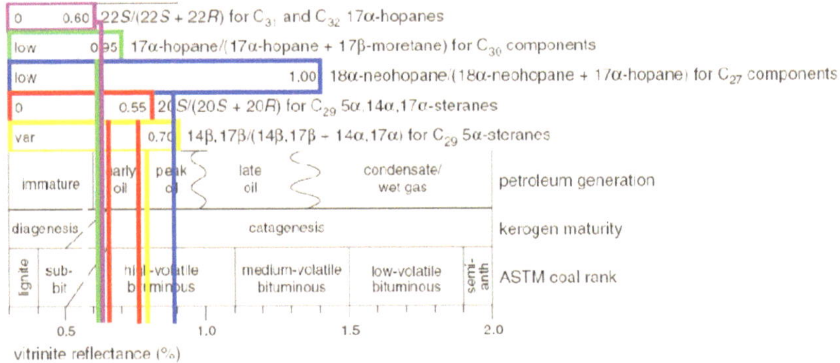

Fig. 14: Maturity parameters 2S/22S+22R C31 and C32 17 α hopanes (purple), 17α hopane/ 17α Hopane+17β hopane (C_{30}) (green), Moretane/hopane (light blue), 20S/(20S+20R) (red) and 14 β 17 β /14 β 17 β +14 α 17 α (C_{29}) (yellow) compared with the vitrinite reflectance (R_o%) and the stages of oil generation. Modified after Peters et al. 2005.

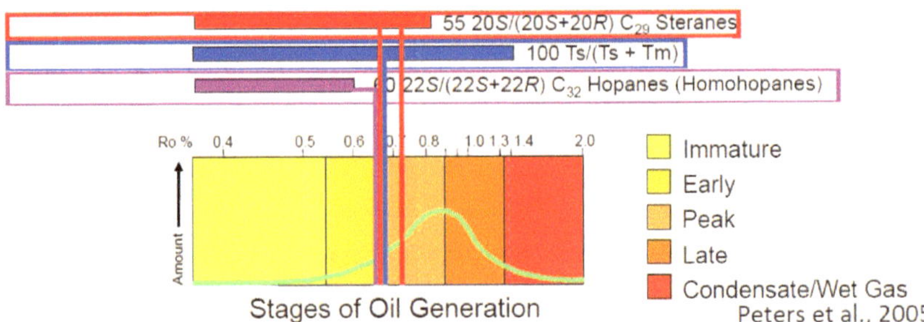

Fig. 15: Maturity parameters 20S/(20S+20R) (red), TS/TS+TM (blue) and 22S/22S+22R (purple) compared with the vitrinite reflectance (R_o%) and the stages of oil generation. Modified after Peters et al. 2005.

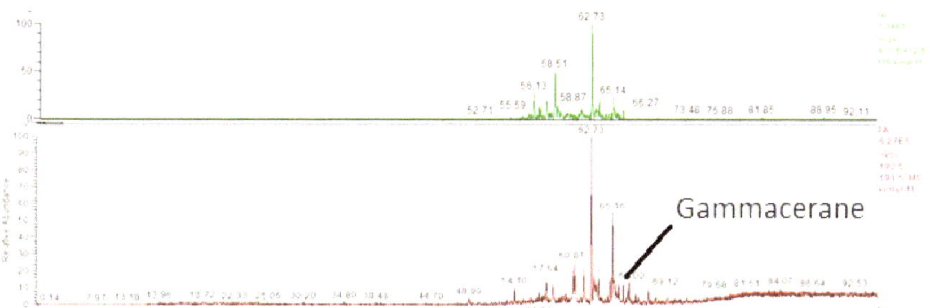

Fig. 16: M/Z 412 with gammacerane/oleanane intensities and retention time indicated by numbers on peaks (green) and m/z 191 (red) and a candidate for a possible peak revealing gammacerane.

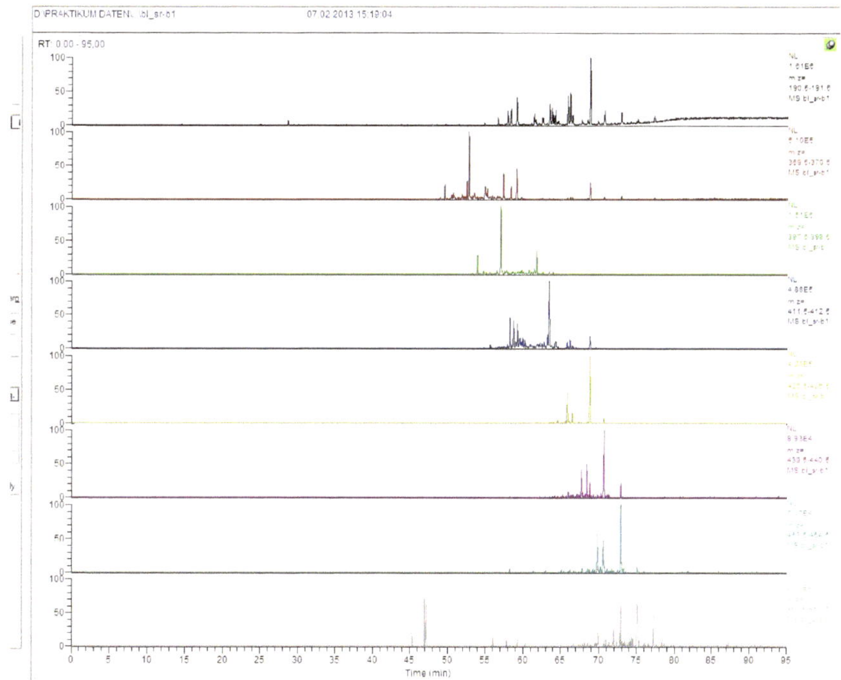

Fig.17: Mass Chromatograms of the Blue Lias saturated fraction, filtered by the characteristic fragment of Hopane (m/z=191) and the most prominent Norhopanes, Hopanes and Homohopanes.

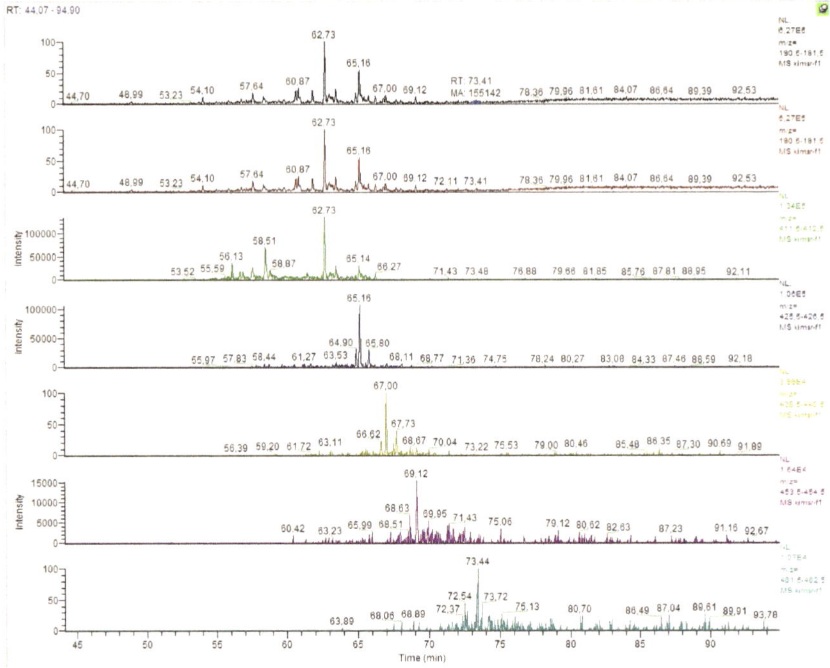

Fig.18: Mass chromatograms of the Schwedeneck saturated fraction, filtered by the characteristic fragment of Hopane (m/z=191) and the most prominent Norhopanes, Hopanes and Homohopanes.

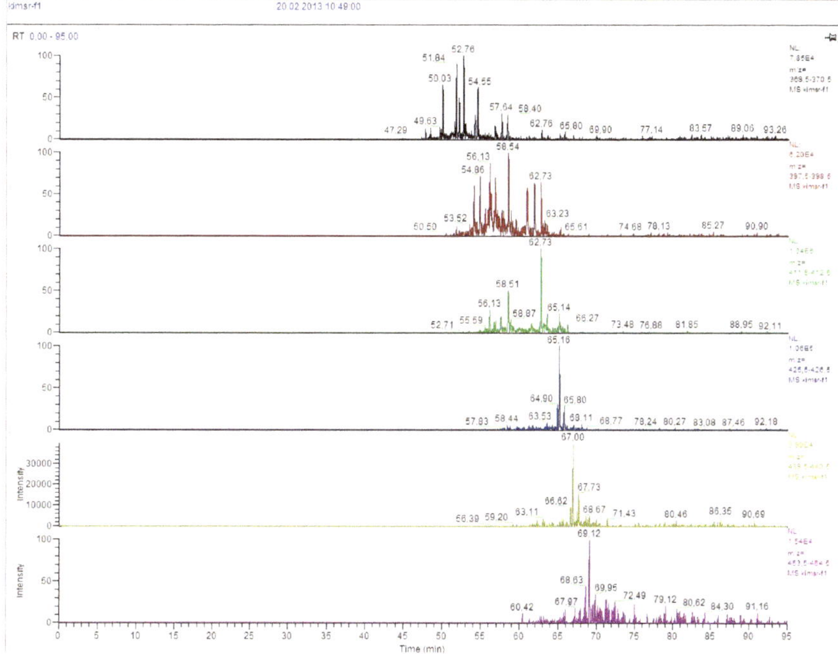

Fig.19: Mass chromatogram of the Kimmeridge Clay 1 saturated fraction, filtered by the characteristic fragment of Hopane (m/z=191) and the most prominent Norhopanes, Hopanes and Homohopanes.

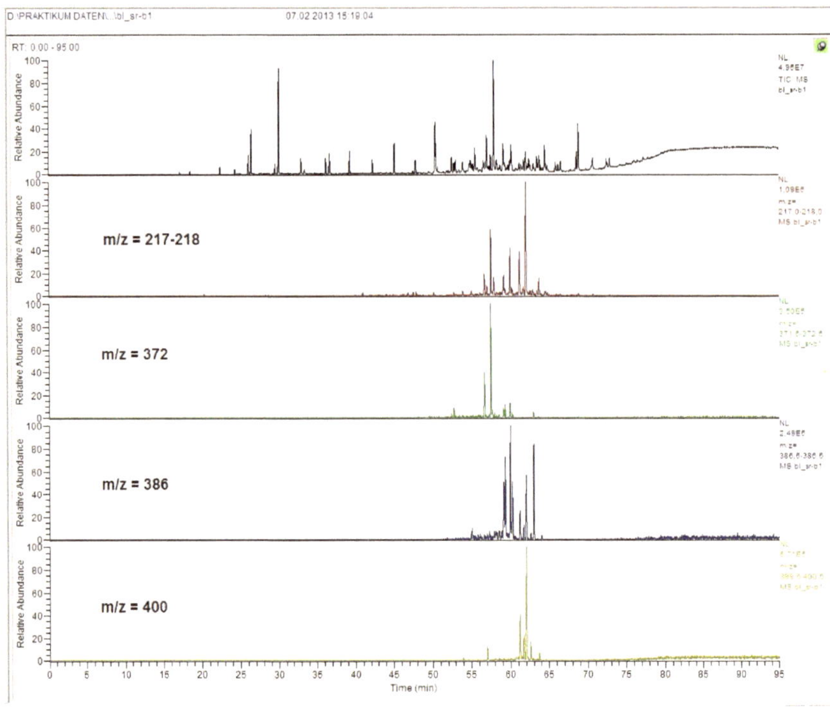

Fig.20: Mass chromatograms of the Blue Lias saturated fraction, filtered by the characteristic fragment of Sterane (m/z=217) and the three most prominent Steranes, Cholestane, Ergostane and Stigmastane.

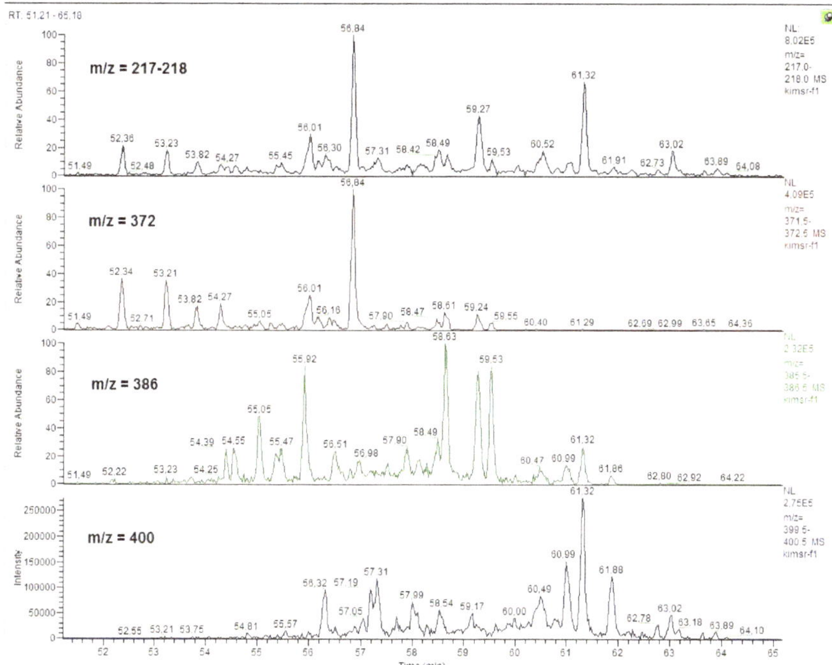

Fig.21: Mass chromatograms of the saturated fraction of the Schwedeneck Oil, filtered by the characteristic fragment of Sterane (m/z=217) and the three most prominent Steranes, Cholistane, Ergostane and Stigmastane.

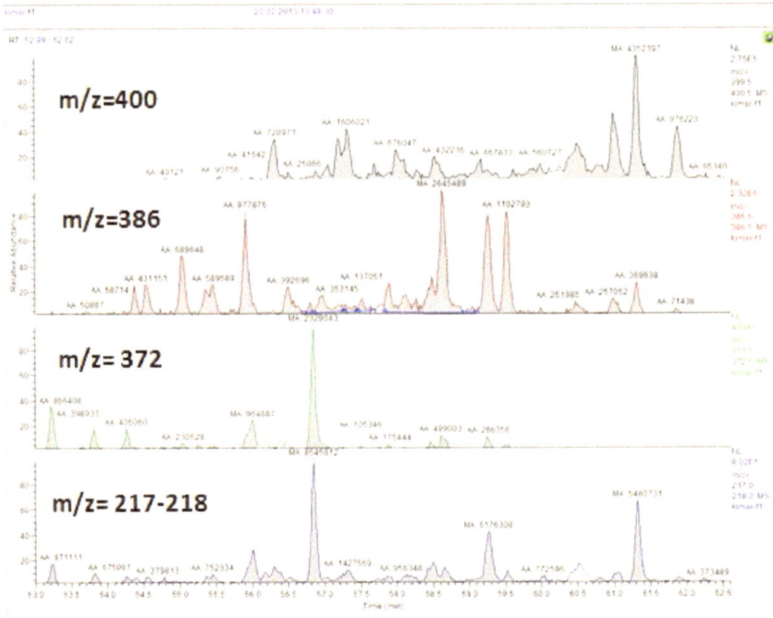

Fig.22: Mass chromatograms of the saturated fraction of the Kimmeridge Clay 1, filtered by the characteristic fragment of Sterane (m/z=217) and the three most prominent Steranes, Cholistane, Ergostane and Stigmastane.